NEW HOLLAND TRACTORS

Jonathan Whitlam

AMBERLEY

First published 2018

Amberley Publishing
The Hill, Stroud,
Gloucestershire, GL5 4EP

www.amberley-books.com

ISBN: 978 1 4456 7767 5 (print)
ISBN: 978 1 4456 7768 2 (ebook)

British Library Cataloguing in Publication Data.
A catalogue record for this book is available from the British Library.

Typeset in 10pt on 13pt Celeste.
Typesetting by Amberley Publishing.
Printed in the UK.

Contents

Introduction

The name New Holland is synonymous with tractors and farm machinery today as part of the giant CNH Industrial group under Fiat ownership. But the story of the farm machinery firm of New Holland goes back a long way, to 1895 in fact, when Abram Zimmerman began a repair shop in the town of New Holland, Pennsylvania, in the USA.

The New Holland Machine Company did well and by 1900 was also manufacturing machinery such as stationary engines and milling equipment. In 1903 the name was changed to the New Holland Machine Company and quarrying machinery was added to the growing list of products.

It was not until 1940, however, that farm machinery really began to become a core part of the business, when the firm began production of a baler invented by a local farmer. Sold as the New Holland Automaton, this was the first truly successful pick-up baler and was also a self-tying machine. Balers would, from now on, form an intrinsic and important part of the business, with a whole range of machines being offered.

In 1947 the company was taken over by the Sperry Rand Corporation, which saw the addition of new machinery, such as forage harvesters and manure spreaders, to its list of products, as well as the opening of new factories. There was now no stopping New Holland as it continued to add more and more farm machinery products to its line, including bale wagons and self-propelled forage harvesters.

Combine harvesters joined the line-up in 1964 when Sperry New Holland, as the firm was now known, bought the major share of Claeys, a company based in Belgium that was one of the largest producers of combines in Europe. This led to New Holland becoming a world presence in the combine harvester market, including, later, a range of landmark rotary models.

The takeover of the Claeys business, which was primarily now one of combine manufacture, was also the first time that New Holland came into contact with a tractor builder, as for a short time Claeys had manufactured its own tractors in Belgium.

Then, in 1986, the Ford Motor Company bought New Holland from Sperry with the result that the farm machinery and tractor businesses were joined together under the name of Ford New Holland. This new company was a great match with Ford, bringing a very popular range of tractors to the mix and allowing the company to become a 'full-liner', to use the American parlance.

This was the position until 1991 when Fiat bought Ford New Holland and then merged it with its existing Fiatagri brand, which also included tractors, combine harvesters and self-propelled forage harvesters. A spell of both tractor ranges being sold separately would give way to both the Ford and Fiatagri brands disappearing under one name: that of New Holland.

Today, blue New Holland tractors are a worldwide force in agriculture along with combine harvesters and forage harvesters. This book is the story of how the two tractor brands were united as one and how the best DNA of both was integrated into a new product that has sold the world over. As we shall see, the mergers and takeovers were not over, but with every change that involved the creation of today's CNH, more benefits have come to the New Holland tractor range, and they have then shaped the modern machines of today that bear that famous name.

For 123 years the New Holland name has been known for making farm machinery and today that name is proudly worn by a whole range of farm equipment, including some of the most sophisticated tractors around. This book is their story.

The Fordson Model F first appeared in 1917 and was the first tractor to be produced by Henry Ford.

Fiat brought out the 702 tractor in 1919 and this was their first production tractor.

The Fiat 640 was the best-selling model in this Italian tractor maker's range for many years. (Photograph: Sascha Jussen)

Ford tractors were built in a factory in Basildon, Essex, from 1964 and had evolved into machines like this 5600 by the middle of the 1970s.

Fiat tractors for the 1970s included this large 1380 model, wearing the distinctive orange livery of the time and also featuring the square bonnet styling and distinctive cab design.

Right: New Holland were primarily building balers and grassland equipment, but the addition of Claeys brought combine harvesters into the fold.

Below: The Fiat 90 Series was extremely successful and became the Fiatagri 90 Series with a terracotta colour scheme. This 140-90DT is a 140 hp six-cylinder tractor. (Photograph: Sascha Jussen)

Launched in 1987, the Ford 7810 became the best selling Ford tractor and was still in production in 1991 when Fiat bought Ford New Holland.

The Ford 8830, built in Belgium, was the largest rigid tractor built by Ford New Holland and was produced up to 1994 under Fiat ownership.

Ford New Holland had a very strong presence in the global combine harvester market, with models such as this TX36 proving very successful.

Laverda combines were part of Fiatagri and machines such as this 3750 were popular in many markets. The New Holland and Laverda combine ranges would be merged in much the same way as the tractors would eventually be.

CHAPTER 1

New Name in Tractors

Joining two very different ranges of tractors together as one brand is not an easy task and Fiat took the process slowly at first. Fiat themselves had a long history of tractor building, going right back to 1919, and had, by the 1970s, become one of the best-selling tractor brands in Europe with a whole range of tractors made distinctive by their unique design, unusual style of cab and their bright orange paintwork. The range included three-, four- and six-cylinder tractors, as well as crawlers, catering for all areas of the agricultural market.

In 1983 the Fiatagri brand was created, joining together the Fiat tractor business with the Laverda combine harvester and Hesston baler and forage harvester lines. This saw a new terracotta livery for all the tractors and machinery in the group, the tractors losing the earlier orange colour scheme for the somewhat drabber brown, but new features included more powerful machines, powershift transmissions and an even wider choice of models. In 1990 the new Fiatagri Winner range of six-cylinder tractors was launched and was soon joined by the smaller 94 Series, although several models of the older 90 Series tractors remained in production, due mostly to the fact they were so popular with so many end users. The new Winner range was an advanced product, not only including all-new cabs and engines, but also updated transmission options. It is interesting to consider what this range might have evolved into if the merger of Fiat and Ford New Holland had not gone ahead!

When Fiat bought Ford New Holland in 1991, which is what it amounted to even though it was claimed to be a merger at the time, they not only acquired another range of combine harvesters, balers and forage equipment, but also became the owners of a line of very popular Ford tractors, the Versatile four-wheel drive articulated giant tractors acquired by Ford New Holland in 1987 and factories building tractors in Canada, Belgium and the UK to add to those it already had in Italy. Astutely, Fiat negotiated permission to retain the Ford name and logo for a certain number of years after the takeover, and this was crucial to the early acceptance of the new arrangements.

The reason for this was that Ford New Holland was just about to launch a completely new range of tractors to replace all the mid-range models currently in production and which Ford had been developing since 1985. By retaining the Ford name, the Series 40 tractors could be launched in 1991 with a sense of continuity intact. The new range stretched from the 75 hp 5640 four-cylinder model up to the six-cylinder 120 hp 8340, with six models in all featuring brand-new cabs, transmissions, hydraulic systems and engines.

Built in the Basildon factory in Essex, the option of a semi-powershift transmission was a new step for the Ford tractor line, as was the increased use of electronically controlled three-point linkage. In fact, the level of technical sophistication included in the top of the range Series 40 models was a huge step up from what had gone before and had been designed to take the Ford tractor through the 1990s. Instead, they would prove to be the last all-Ford tractors to be produced, ending a dynasty that stretched right back to 1917.

Both the Ford New Holland and Fiatagri brands remained unchanged for the next few years, and remained completely independent of each other, but a new tractor range launched in 1994 saw the first indications of what would soon be coming, not only in their design but also in the way in which they were branded. Built in the Versatile factory in Winnipeg, Canada, the new high horsepower range replaced the Ford six-cylinder 30 Series and the largest Fiatagri 90 Series models, and included much more powerful tractors than had featured in either range previously.

At the heart of the new high-power machines was a full powershift transmission sourced from Funk and a range of six-cylinder PowerStar engines from 170 hp to 240 hp over four models. Three computer systems were included to control the tractors' functions and a sleek, sloping bonnet covered the engine compartment and could be lifted in one piece for maintenance, giving previously unheard of ease of access. A neat option introduced on these machines was the SuperSteer front axle, which was of a brand new concept and pivoted to give a previously unobtainable tight turning circle of 65 degrees. The big news though was that all four tractors in this range would be sold in two identities – in blue as the Ford Series 70 and in terracotta as the Fiatagri G Series. Otherwise, the tractors were identical.

This was the first instance of a new tractor range being launched in the blue and brown livery of the two brands and was the first step to a unified tractor range. To begin with, the Ford oval was used on the front of the Series 70 tractors and the Fiatagri leaf emblem on the G Series, but the Ford oval logo was only allowed to be used for a few years and so the Fiatagri leaf emblem was changed to blue and this was used on the front of the blue Ford-badged tractors, including the Series 40 machines, which received a blue cab roof and grey chassis in 1994.

From 1996 the Series 70 and G Series tractors were branded as New Holland with the Ford and Fiatagri names relegated to small lettering on the bonnet sides, the blue New Holland leaf emblem being used on the front of both colours, as it had now become the New Holland logo. The Series 40 tractors were treated in the same way, with New Holland branding instead of Ford.

It was in 1996 that the first truly new tractors with both Ford and Fiat DNA were launched and they took the form of two ranges, one comprising new mid-range six-cylinder tractors and the other of smaller three- and four-cylinder machines, which were basically of Fiatagri parentage.

The smaller tractors were in fact actually revamped Fiatagri 94 Series tractors, with new cabs and sloping bonnets, and were sold in blue as the New Holland Series 35 and in terracotta as the New Holland L Series. All were initially built at the Fiatagri Jesi factory in Italy. The range spanned from the 60 hp 4635/L60 to the 95 hp 7635/L95 and included a one-piece sloping bonnet that could be raised for access to the engine for maintenance. The Ford and Fiatagri names were still carried but were much reduced in size.

The same style of sloping bonnet, although less raked, was used on the bigger new tractors, the New Holland Series 60 and M Series. Again, both were identical except for colour schemes and this range of tractors was much more sophisticated than the smaller 35/L Series, with all the best parts of the Fiatagri Winner and Ford Series 40 tractors being brought together to produce a new, class-leading tractor range, with six-cylinder models from the 100 hp 8160/M100 to the 160 hp 8560/M160, with four models in total. Power came from 7.5 litre PowerStar engines and all were built at the former Ford factory in Basildon, England. These proved to be an important introduction as they not only set the stage for future models, but also cemented the Fiatagri and Ford brands as New Holland, and also proved popular machines that were well and truly capable of taking on the competition.

These two new introductions of 1996 set the stage for what was to come and, along with the 70/G Series, set the trend for future New Holland tractor ranges. The 60/M Series, however, were the truly landmark tractors, combining seamlessly as they did both Fiatagri and Ford design and technology in a modern package. The range was also extremely successful sales-wise in Europe, the USA and other parts of the world, often with no cab fitted. The foundation of the New Holland tractor range was being laid.

Despite Fiat now owning the majority share of Ford New Holland, the Series 40 was launched in 1991 as a Ford product. This 6640 SLE was a full specification four-cylinder 85 hp tractor.

Inside the cab of the Series 40 and the layout was very different to the older Ford tractors, with the semi-powershift gearbox controls to the left and hydraulic and power take-off controls to the right.

The 100 hp six-cylinder 7840 replaced the popular 7810 model. This is the more basic SL version, which came with a manual transmission and hydraulic system. Dual Power was added later, giving double the number of gear ratios.

This slightly careworn 8340 is an example of the Series 40 flagship model. Originally rated at 120 hp, this was soon increased to 125 hp and was only available in SLE form.

Smallest of the Series 70 was the 170 hp 8670, which was also sold in terracotta as the Fiatagri G170. Sleek styling thanks to a one-piece tilting hood made these tractors look very modern, and this was further enhanced by the amount of electronic control and monitoring systems fitted. When launched in 1994 the Ford name was still prominent. (Photograph: Sascha Jussen)

A Fiatagri G170 showing what a difference colour can make to a tractor. This is a later example with the blue New Holland badge on the front. (Photograph: Kim Parks)

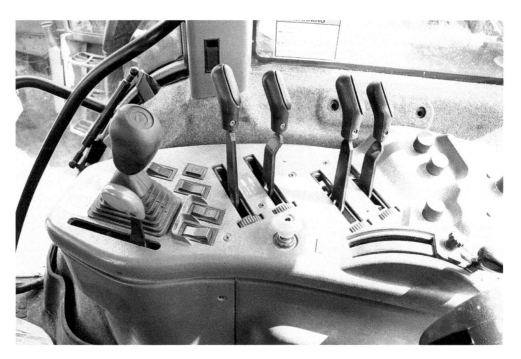

Inside the Series 70/G Series the full powershift transmission was controlled by a small gear lever and all the other main controls were positioned conveniently on a console to the right of the driver's seat.

This 7840 is an example of an interim model in that it still features Ford branding but now has a blue cab roof and grey chassis. The Ford oval has vanished from the front in favour of the blue New Holland 'leaf' logo.

Three out of four of the New Holland Series 70 range, left to right: 210 hp 8870, 190 hp 8770 and 170 hp 8670. The Ford name is now very small, under the model number, and New Holland is the prominent branding.

The last overhaul of the Series 40 range saw them become a New Holland tractor with only a very small Ford badge under the model number. This 95 hp four-cylinder 7740 is hedge cutting in Suffolk.

Despite being twenty years old, this New Holland 8340 still looks pretty much new as it takes a break from rowing up grass in Lancashire.

Remaining popular right to the very end, the 7840 was the last Series 40 tractor to remain in production, right up to 1998.

The 95 hp four-cylinder 7635 was the largest in the New Holland Series 35, launched in 1996. Based largely on the Fiatagri 94 Series, these tractors were ideal for livestock and mixed farms, and many were equipped with front loaders.

Fiatagri versions of the Series 35 were known as the L Series, and this L85, planting lettuces in Germany, was equivalent to the blue 6635. (Photograph: Sascha Jussen)

At 65 hp the four-cylinder 4835 was the second smallest in the range. Note the steeply sloping bonnet fitted to all this range of tractors.

In 1991, just before Fiat acquired Ford New Holland, Fiatagri launched the advanced Winner range of six-cylinder tractors with an updated range, of which this 115 hp F115 model was a part, appearing in 1993.

Both Fiatagri Winner and Ford Series 40 technology was used to produce the Series 60/M Series in 1996, making these the first true New Holland tractors to appear. This 100 hp 8160 was used by an East Sussex contractor.

The terracotta version of the 8160 was the M100. As with the Series 70/G Series, the only difference was the colour. (Photograph: Sascha Jussen)

Seen power harrowing a seedbed in Norfolk, the 8260 was a six-cylinder 115 hp tractor. The sloping one-piece bonnet gave excellent access to the engine as it tilted upwards out of the way.

Built at the factory in Basildon, the Series 60/M Series also used new engines built in the same facility. In the 8360 model the PowerStar 7.5 litre engine was rated at 135 hp.

A flagship 160 hp M160 collects maize from a Krone forage harvester in Germany. The M160/8560 was the most powerful tractor to be built at Basildon so far. (Photograph: Sascha Jussen)

Above: This 4835 shows off the new styling of the New Holland 35 Series at the 1996 Grassland '96 event in Warwickshire.

Right: Inside the cab of a well-worked 7635, showing the flat floor and easy access to the comfortable seat.

The 35/L Series tractors were based on the earlier Fiatagri 94 Series, such as this 88-94, but new cabs and bonnets made them look like a completely new tractor.

A 95 hp New Holland L95 showing how different the L Series looked to its 94 Series ancestors.

A close-up of the bonnet of the 8260 showing both how it slopes to give excellent forwards visibility and the absence of an exhaust stack, which is now mounted to the side of the cab.

An 8360 at work, applying fertiliser in Kent ahead of potato planting. (Photograph: Kim Parks)

A flagship 8560 model at rest during the winter.

The terracotta flagship M160 at work on a planter in Europe. (Photograph: Sascha Jussen)

The 60 Series contributed major parts, including the engine, for the unique TV140 built in Canada which continued the tradition of a bi-directional tractor begun by Versatile before the takeover by Ford New Holland. (Photograph: Paul Reeve)

A New Holland TV140 on demonstration in the UK. Not many were sold into Europe, but it was popular in North America.

CHAPTER 2

Blue Letter Day

The launch of the 60/M Series in 1996 effectively replaced the six-cylinder models of the earlier Series 40 range and, in 1997, a new range of sloping bonnet tractors called the TS series replaced the smaller tractors. This marked a new development as the letter system used previously on the Fiatagri range was expanded to take in the blue machines as well, but in a new format. The TS machines were built at Basildon and were based on the Series 40 platform with three four-cylinder models ranging from the 80 hp TS90 to the 100 hp TS110. The new model numbers and the steeply raked bonnets made these tractors look very different from their predecessors, even though they shared the same cabs and mechanicals, the sloping bonnet being something of a fad at the time and something adopted by several manufacturers on certain models or, by some such as Deutz-Fahr, across the entire range. The benefits included better forward visibility when using front-mounted implements and front loaders.

The extremely popular 7840 remained the sole survivor of the Series 40, and the Ford line, with its six-cylinder 100 hp engine providing a lighter package compared to the 8160 of the same power. All that changed in 1998 when the six-cylinder 100 hp TS115 finally replaced it and, in keeping with its predecessors, went on to become the best seller of the entire TS range.

Fiat had always had a strong market presence in the smaller orchard type tractors, which were built to negotiate narrow orchards and vineyards. The original Fiatagri models were replaced in 1997 by the TN range, which was produced in both blue and terracotta. Many different versions were offered between 65 and 88 hp and there was even the option of the SuperSteer front axle. The TN range continued to expand and develop over the years, bringing ever more sophistication to this sector of the tractor market and the amount of differing specifications and features on the TN series was astonishing, with models to suit every possible need in the lower horsepower sector.

In 1998 the 35/L Series was replaced by the TL range, with four models from the 65 hp TL70 to the 95 hp TL100, and continued to be offered in both blue and terracotta, as well as featuring the sloping bonnet design of their predecessors.

The following year saw the 60/M Series bow out in favour of the similar looking TM range, which was once again produced in both colour schemes and still built at Basildon.

They might have looked similar to the older models, but the TM range of six-cylinder tractors, from the 115 hp TM115 to the 165 hp TM165, were packed full of new sophisticated innovations, including the option of the TerraGlide suspended front axle, SuperSteer front axle and cab suspension. New semi-powershift and full powershift transmissions were also offered. A TM165 Ultra model was also produced, with full specification including front axle and cab suspension, as a top of the range model.

The TM range was just what was needed in the mid-size range, with its advanced features and long list of options. Tractors from all the leading manufacturers were becoming ever more sophisticated at this time and so the TM range had a lot resting on its shoulders. It met expectations in no small measure.

In 2002 the range was updated with new model numbers and larger tractors. The range now stretched from the 124 hp TM120 to the 154 hp TM155, but two larger frame tractors took the power up the scale even further, the largest of which was the 194 hp TM190 – the largest tractor so far built at the Basildon factory – along with its slightly smaller sister, TM175. These larger two models brought the TM into a new power bracket and filled a gap in the range between the TM series and the much larger models built in the USA. They also featured such items as a full powershift transmission and the combination of more power in a relatively small package meant that they soon proved to be a popular choice among farmers and contractors. Indeed, they were the first of what would be a growing trend in later years as the thirst for power continued to grow.

Later versions of the TM range also benefitted from a new, brighter bonnet decal from 2004 to bring them in line with other new tractor models launched since the turn of the century. Building on the success of the 60/M Series, the TM range kept the New Holland name at the forefront of tractor technology and was hugely successful.

When it was time for most of the Series 40 tractors to be replaced in 1997, the new tractors not only featured steeply sloping bonnets but were also the first of a new model numbering system. This TS90, an 80 hp four-cylinder machine, was the smallest.

The TS100 was originally the largest, with 100 hp from its turbocharged four-cylinder engine. Based on the Series 40 platform, these tractors were also built at Basildon.

When the ever-popular 7840 bowed out in 1998, it was replaced by the TS115, the only six-cylinder member of the TS range.

Inside the cab of the TS115, and they would be familiar to any user of the Series 40 machines, but the layout had changed quite a bit.

The Italian-built TN range was a very diverse one, with many specialist models and features. The TN90F was a powerful performer for its small size and was ideal for work on fruit farms, as here in Germany. (Photograph: Sascha Jussen)

The one-piece bonnet of the TN range hinged backwards to gain access to the engine, in this case a four-cylinder 88 hp unit. (Photograph: Sascha Jussen)

The Series 35/L Series only lasted a couple of years before being replaced by the similar TL range in 1998. Electronics were now included though, the digital instrumentation being ideal for jobs such as spraying. (Photograph: Sascha Jussen)

The four-cylinder 95 hp TL100 was the largest model in the TL range and could also be equipped with a front linkage to increase its versatility. (Photograph: Sascha Jussen)

In 1999 the Series 60/M Series bowed out in favour of the new TM range. Similar in many respects, the TM range featured a higher level of sophistication and a greater choice of options. This TM150 is collecting maize from a New Holland forage harvester. (Photograph: Sascha Jussen)

At 150 hp, the TM150 was the second from largest in the new TM range.

The biggest was the 165 hp TM165 flagship model, which is seen here rolling a seedbed in Wiltshire. (Photograph: Kim Parks)

The TM165 could be equipped with a range of different options, including a front linkage. The Ultra version included both front axle and cab suspension.

A significant upgrade of the TM range in 2002 saw the introduction of new models, including the 144 hp TM140.

The 154 hp TM155, shown here with the later brighter bonnet decals, was now the largest in the standard build TM range.

In certain markets the TM range was still produced in terracotta livery, as shown by this TM155 model in Germany. Soon, though, all New Holland tractors would be blue. (Photograph: Sascha Jussen)

Two larger build TM tractors arrived in 2002 in the shape of the TM175 and TM190. The TM175 had a power rating of 177 hp and came with a full powershift transmission.

Inside the cab of a TM175 shows the main controls, including the lever used for controlling the Power Command transmission.

A full front view of the TS115, showing the new distinctive features of the new six-cylinder addition to the TS range.

A steeply sloping bonnet was a striking feature of all the TS range.

The cab used on the TS range was carried over from the previous 40 Series models, although there was a slight change to the internal layout.

Above: The TM175 was a large tractor for its time and had an imposing side profile.

Left: The view up into the cab of the TM175.

CHAPTER 3

Artic Power

The story of large, powerful, articulated steer four-wheel drive tractors is without doubt a North American one. The idea of using a pivot steer system to produce a high-powered agricultural tractor was pioneered in the USA by Wagner in the 1950s, but the Steiger brothers were not far behind. Also based in North America, the Steiger tractors, painted in their light green livery, soon became one of the best-selling articulated tractors in the world. One of their main competitors was Versatile, based in Canada, who also produced large articulated four-wheel drive tractors.

From 1977 Ford sold a range of articulated four-wheel drive tractors produced in Ford branding by Steiger, and two of these, the FW-30 and FW-60, were remarkably successful in the UK. Fiat, meanwhile, entered into a similar arrangement with Versatile to produce a range of Fiat-badged monster tractors for sale in Europe.

Gradually the sales of Ford FW tractors dwindled, and when Case IH bought Steiger in 1986, Ford New Holland bought Versatile to enable it to continue to offer high-powered articulated tractors in its range. Soon, Versatile tractors appeared in blue livery with Ford branding and a whole new range arrived in 1994 in the shape of the Series 80, with new cabs and styling. These wore the New Holland leaf emblem on the front of the bonnets but had the Ford and Versatile names down the sides. With many new features the new range looked very striking, not least because of the huge size of the bonnets that concealed the powerful six-cylinder engines used in these massive machines.

By 1997 these tractors had become the 82 Series and the New Holland name replaced that of Ford. In 1999 the 84 Series arrived and the Versatile name was dropped for the first time in favour of just New Holland. Biggest of these huge machines was the 425 hp 9884 leviathan, a simply huge tractor with massive lugging ability.

Big changes in the world of farm machinery would see the 84 Series being a short-lived range however. It was in this year that Fiat bought Case IH, a global brand, which produced a complete range of tractors and combine harvesters and also included the Steiger articulated four-wheel drive tractors in its portfolio. The result of this huge merger was the creation of the CNH group and this would have big ramifications for the future of the New Holland brand.

The immediate problem the merger faced was complying with international trade monopoly laws, the result of which was that the new company had to divest itself of the Versatile factory in Canada and the Case IH factory in Doncaster, England, and all the tractor models built in them. For New Holland this meant losing the 70/G Series and the 84 Series, plus the unusual bi-directional tractor also built there. It would also lead to red Case IH tractors being produced in New Holland factories and vice versa.

The first example of 'badge engineering' was the New Holland TJ Series, which was based on the Case IH Steiger STX articulated four-wheel drive tractors ranging from 275 to 440 hp and built at Fargo in North Dakota in the USA. This would be the start of a full integration of both the New Holland and Case IH product lines only eight years after the Ford New Holland and Fiatagri merger, and would ensure the future of blue-painted articulated monster tractors.

The Versatile factory in Canada and the 70/G Series and 84 Series were bought by Buhler, who carried on producing the tractors in their own colours and branding, as well as producing them for New Holland.

Fiat by contrast turned to Versatile to produce a range of Fiat-badged articulated high horsepower tractors at around the same time.

Ford turned to Steiger to produce the FW Series from 1977. The FW-60 was the largest at 335 hp. (Photograph: Kim Parks)

When Ford New Holland bought the Versatile business it did not take long before the Canadian-built tractors were badged as Ford and painted blue, as demonstrated by this 946 tractor at work in Leicestershire.

Left: A new design of Versatile tractor arrived in 1994 with the introduction of the Series 80 monster tractors. Looking imposing when brand new and on display at the Royal Norfolk Show, the 9680 was powered by a 350 hp six-cylinder Cummins engine.

Below: The Series 82 giant tractors arrived in 1997 and carried on the design of the earlier range. Power was up across the range, the 9682 being a 360 hp tractor.

Large 'super single' tyres fitted all round help put the power of the 9682 down to the ground but without the added width of dual wheels.

In 1999 the Series 84 articulated tractors arrived but they had a very short life, at least in blue. With dual wheels all round, this 360 hp 9684 looks very imposing as it powers across a Kent field.

From the rear the 9684 looks just as imposing as it pulls a Simba cultivation train incorporating oilseed rape stubble.

CHAPTER 4

New Breed

Buhler Versatile continued to produce the updated 'A' Series of the 70/G range for New Holland until a new replacement was ready. This was the TG Series, which was launched in 2002 and, like the TJ machines, was actually based on a Case IH design and was a continuation of the badge engineering exercise.

This time it was the big six-cylinder Case IH Magnum range that provided the platform for the new blue machines, but there were several changes made, not least in the positioning of the engine itself, which was placed further back to allow room for the SuperSteer front axle. The Magnum cab was kept but a new roof was fitted and new distinctive bonnet styling made the TG range look very impressive and purposeful.

Three models made up the range, from the 246 hp TG230 to the 311 hp TG285, and all included a sophisticated transmission system including automatic modes for various tasks and a parking mode. The result was a brand-new range that actually had little in common with the Magnum it was based on, and thus the TG was certainly a New Holland product through and through despite its origins.

In many ways the TG range was the first sign of a major change in direction for the New Holland line-up, and the styling cues it initiated would soon show in further new model introductions, the first of which was the landmark TSA range that appeared in 2003 and replaced the existing TS tractors.

The most significant thing about the TSA range was the first use of a brand-new cab design, called the Horizon cab by New Holland, and it sat well with bonnet styling that shadowed the bigger TG range. Five models were produced using both four- and six-cylinder engines built by Iveco, another Fiat company, and ranged from the 101 hp TS100A to the flagship six-cylinder 136 hp TS135A. All were built in Basildon and the same platform was used to produce the Case IH Maxxum MXU range.

The TL line of tractors were also updated in 2004 to bring them more in line with the TSA Series, with new bonnet styling and four-cylinder engines, varying from the 70 hp TL70A up to the 100 hp TL100A. All were built in Italy and retained the original cab used first back in 1996 on the 35/L Series.

Gradually all New Holland tractors were painted blue, the old Fiatagri terracotta livery being finally abandoned by the beginning of the twenty-first century. This did not mean that all the old Fiatagri designs disappeared altogether though, as the very successful 100-90 and 110-90 models carried on right up to 2005 and also wore New Holland branding and blue paint. The last of these models carried a 'Tradition' badge on their cabs and were certainly the end of an era, being the very last tractors of true Fiat design to be built. It is quite ironic that the Fiatagri Winner range of 1990 was intended to replace these very models from the 90 Series, but the popularity of the old stagers just made it impossible to delete them from the sales lists and even all the sophisticated tractors introduced throughout the 1990s and early years of the twenty-first century did not supplant them. In the end it was emission regulations that spelt the death knell for them when the engines they used could no longer be made to fall in line with the legal requirements.

Back in 1996 Case IH had bought the Steyr tractor business, based at St Valentin in Austria. Steyr had been producing tractors since the 1940s and was also a trendsetter in tractor design, producing some unusual machines, such as reverse drive tractors. They had also been developing a constantly variable transmission system, which gave stepless control of the whole speed range, and were only just beaten to the marketplace by Fendt in the 1990s. When Case IH took over, the Steyr identity survived and they continued to produce tractors in their distinctive red and white livery, but some models were also badged as Case IH and these soon included the CVT models, badged as CVX.

With the merger of Case IH and New Holland it was perhaps inevitable that Steyr would eventually have an impact on the blue tractor line and this finally happened in 2005 when the New Holland TVT line was introduced, featuring machines from the 139 hp TVT135 to the 193 hp TVT195 and all equipped with the CVT tranmisssion. They were built at the Steyr plant in Austria and were simply Steyr tractors painted blue. As such they were unique in the New Holland stable and were powered by Sisu six-cylinder engines from Finland.

Of course the arrangement worked both ways, and the standard Steyr models also benefitted from sharing a common platform with the New Holland and Case IH product range. New Holland tractors were now being built in the USA, England, Italy and Austria and had become truly a global brand.

In 2007 a revamped T7500 range from Austria replaced the original TVT models and offered powers of between 174 and 216 hp. Sales were slow, however, and this range was only offered up to 2010. Soon constantly variable transmissions, which had become an extremely popular choice for users, would be an integral option on the New Holland tractor range proper.

In 2001 the Series 70A range appeared as an upgrade to the original models but was now built by Buhler Versatile for New Holland following the sale of the Canadian tractor factory. Largest was the 8970A.

In order to replace the Series 70A range New Holland introduced the TG Series in 2002, based on the Case IH Magnum platform built in Racine in the USA but featuring a shorter bonnet to accommodate the Super Steer axle brought over from the earlier range.

Largest of the TG range was the 311 hp TG285, which drew its power from a six-cylinder 8.3 litre turbocharged and intercooled engine.

The TSA range were built in Basildon and included both four- and six-cylinder tractors and bonnets that followed TG styling. The TS115A was a six-cylinder 115 hp machine.

Biggest in the TSA range was the TS135A, with a rated output of 136 hp but a boost feature taking this up to 169 hp at the power take-off.

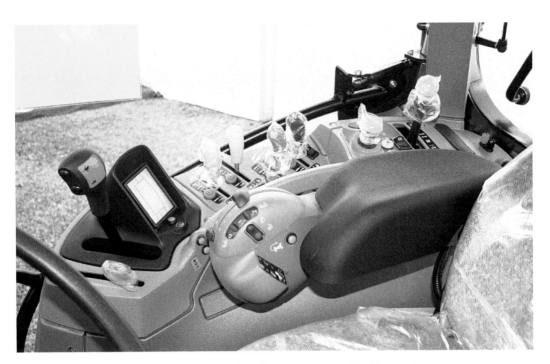

A big feature of the TSA range was the all-new Horizon cab with very large expanses of glass. All the controls were grouped neatly to the right-hand side of the seat and included some built into the seat's armrest itself.

With front loader brackets in place, this TS135A is carting grass for silage in East Sussex.

The TLA range arrived in 2004 and was still built in Italy. Largest was the 100 hp TLA100A.

The main difference on the TLA tractors over the earlier TL range was that the new family of engines fitted were of the same type as those first introduced in the TSA range the previous year.

The 72 hp TL70A was the smallest in the TLA range.

So popular were the 100-90 and 110-90 tractors, which were originally Fiatagri machines, that they remained in production alongside the new blue tractors right up to 2005. The last batches were painted blue with black cabs, as on this 110-90 used by a Suffolk farmer and contractor.

Case IH had a constantly variable transmission option available in their range thanks to badge engineering models from the Steyr range and in 2005 a blue version appeared. The TVT190 was the largest and was powered by a six-cylinder 193 hp Sisu engine.

As can be seen here, the one-piece bonnet of the TVT range could be lifted up vertically to give easy access to the engine bay.

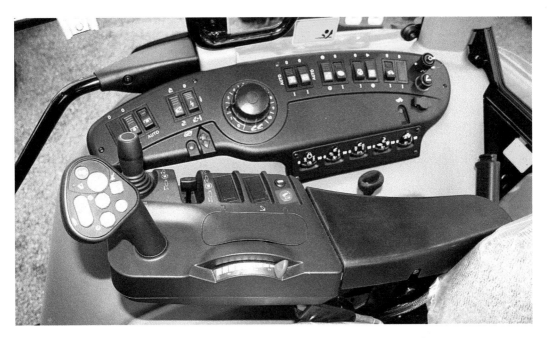

Inside the cab of a TVT190 and the main controls, including those for the CVT transmission, were positioned on the seat's armrest.

In 2007 the TVT range was altered and became the T7500 Series, with the T7550 being the largest with a maximum power output of 216 hp. (Photograph: Sascha Jussen)

CHAPTER 5

The Thousand Effect

The year 2005 also marked the beginning of a new paradigm of New Holland tractor models when the similar looking T8000 line replaced the TG range, beginning a new system of thousand numbering.

One thing that had changed was the fact that power outputs had gone up once again in the ever-increasing search for more power. Still based on the Case IH Magnum platform, and still keeping the set back engine position of its predecessor, the T8000 range included three models, from the 281 hp T8020 up to the 337 hp T8040 and, in 2008, the T8050 arrived boasting an earthshattering 358 hp, taking the range up to what had once been articulated tractor territory.

In 2006 the larger models in the TM line were finally replaced and this was the end of the line for the cabs first seen on the Ford Series 40 models as the new T7000 range were equipped with the more modern Horizon cab. Once again the styling cues of the original TG range were followed and the TM tradition of offering a high level of sophistication and a huge option list was also continued, including such items as TerraLock headland management, FastSteer lock to lock steering, SuperSteer and Terraglide suspension. All were six-cylinder tractors ranging from the 167 hp T7030 up to the 213 hp T7060 and featured a power boost function, providing more power at the power take-off for certain tasks. Like the TM machines before them, the T7000 range was built at Basildon and was exported worldwide, the Case IH version being called the Puma range and based on the same platform but with many detailing changes. These were also built at Basildon, until the bulk of Puma production was switched to the Steyr factory in Austria.

The T7070 arrived in 2009, taking the T7000 series even further up the power scale with a 225 hp rating and power boost up to 251 hp. This model also saw the introduction of a new constantly variable transmission called AutoCommand. In 2010 the T7000 won the 'Tractor of the Year' award and special Blue Power editions of the T7070 and T7060 were produced to celebrate, featuring metallic blue paintwork and chrome-effect detailing.

In 2007 the TSA range was replaced by the T6000 models to bring them in line with the thousand numbering system. The choice offered by the TSA models was enhanced even further with three four-cylinder tractors from the 101 hp T6010 up to the 122 hp T6040 and three six-cylinder models from the 117 hp T6030 to the flagship 155 hp T6080. Various specifications were available, including Elite, Delta and Plus, denoting various specification levels, and both front axle and cab suspension was offered. All in all this proved to be a very comprehensive range, with options to suit every purpose and eventuality.

2009 saw the T6090 arrive, taking the T6000 range up to 165 hp and with a higher build than the other machines, in much the same way as the larger frame TM tractors of several years earlier, although the power output was less. Soon a new T6080 followed, being built to the same dimensions as the T6090, and both machines offered high horsepower but without the weight of the larger T7000 models.

The T3000 range filled a gap in the 35 to 54 hp compact market in 2007 and was actually sourced from Landini of Italy. These were sold in addition to the Boomer range of compacts being built in the USA until this was expanded further and the Boomer line replaced the Italian-built tractors.

At the other end of the scale, the T9000 range took the power of New Holland tractors up to 530 hp on the flagship T9060 model from 2008 with its Cummins six-cylinder 15 litre power plant. Still based on the Case IH Steiger machines, the T9000 line was sold more globally than the TJ range it replaced, including in the UK.

It was also in 2008 that the Italian-built T4000 line replaced the very popular and wide variety of TN models and included standard tread machines as well as narrow variants from 65 to 97 hp. Meanwhile, the T5000 range from Italy replaced the TLA line and looked very similar, although new engines from 76 to 106 hp powered them, with the 113 hp T5070 arriving in 2009. High clearance 'mudder' versions of these tractors were also built.

New engines were at the heart of the T8000 range that replaced the TG Series in 2005, although outwardly they remained pretty similar. The 8030 was a 237 hp tractor with power boost up to 307 hp.

Largest in the new T8000 range was the 337 hp T8040.

The new brighter New Holland decals looked striking on the T8000 range's tall bonnet.

Inside the cab of the T8000 and the full powershift transmission and other main features were all controlled from the armrest console.

In 2008 the T8000 range increased with the addition of a new 358 hp T8050 model, taking the power envelope even further upwards for rigid frame tractors.

The T7000 line introduced in 2006 saw a major introduction of new Basildon-built tractors and more power than ever before. Equipped with the Horizon cab, the T7050 was a 197 hp tractor with boost up to 234 hp.

The controls in the Horizon cab of the T7000 range included armrest-mounted items as well as digital readout.

The largest model in the original T7000 line-up was the T7060, with a maximum of 244 hp.

The T7040 was a very popular model in the new range with its maximum power of 224 hp.

The T7000 looked very different to the TM tractors it replaced, even from the rear, where the new cab and wide rear mudguards are very clearly seen on this T7040.

A T7040 working in Europe topping and lifting sugar beet in one pass. (Photograph: Sascha Jussen)

In 2009 the T7070 was added to the range, with a maximum power of 251 hp. This one is wearing the special metallic blue and chrome finish applied to a number of top specification T7060 and T7070 tractors in 2010.

The TSA range became the T6000 line in 2007 and featured more models and more choices of specification. Smallest was the four-cylinder 101 hp 6010, which is shown here working in an onion crop in Holland. (Photograph: Sascha Jussen)

A 117 hp six-cylinder T6030 taking clods and stones out of a potato seedbed in Suffolk.

The 155 hp T6080 was the largest in the original T6000 line-up.

In 2009 the 165 hp T6090 arrived as the new flagship of the T6000 range, with a higher build than the other models, which was then also adopted by a new version of the 6080.

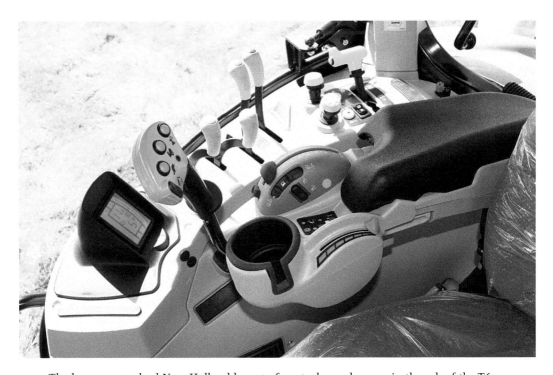

The by now standard New Holland layout of controls can be seen in the cab of the T6090.

Left: Built in Italy by Landini, the T3000 range included the four-cylinder 47.7 hp T3030, which is seen here working in a fruit orchard in Germany. (Photograph: Sascha Jussen)

Below: The Case IH Steiger STX-based TJ range was not imported into the UK, but members of its successor T9000 range certainly were. The 530 hp T9060 was the most powerful.

A 380 hp T9050 cultivating land in Kent. (Photograph: Kim Parks)

This T5060 working with a New Holland 548 round baler in East Sussex was the largest of the T5000 line-up with its four-cylinder 106 hp engine. (Photograph: Kim Parks)

The T4000 range was an extensive one, encompassing all the different variations previously catered for by the TNA line-up. The T4050N was a four-cylinder tractor of 97 hp and was a narrow version for vineyard and orchard use. (Photograph: Sascha Jussen)

The four-wheel drive high clearance version of the 95 hp TD5050 makes an impressive sight as it works in an asparagus crop. (Photograph: Sascha Jussen)

CHAPTER 6

Powerful to a 'T'

Nothing ever stays the same for long in the world of farm machinery and from 2010 a new model numbering system was adopted for the New Holland tractor range. First to appear was the Basildon-built T7 range that replaced the T7000 models, with six-cylinder tractors from the 171 hp T7.170 up to the 269 hp T7.270. All were fitted with new FPT (Fiat Power Train) engines designed to meet Tier 4A emissions targets and included Selective Catalytic Reduction technology using Ad Blue. A huge range of both transmission options and control features was offered and some were even finished with a British flag bonnet transfer!

Blue Power versions were also still offered with full specification and the metallic finish and, in 2014, to celebrate fifty years of tractor production at the Basildon factory, a number of T7 tractors were painted in a special metallic blue and gold livery.

The T8 range also replaced the T8000 tractors in 2010 and this time the forward engine position of the Magnum was retained to give better weight distribution than the earlier tractors had been able to achieve. New aggressive looking bonnet styling made these tractors stand out from the crowd and power was once again on the up with models stretching from the 298 hp T8.300 to the 389 hp T8.390. FPT 8.7 litre engines provided the power driving through a choice of full powershift transmissions.

Similarly, the T9 range took the power envelope even further from 2010 when the new FPT engines under the expansive bonnets provided power from 390 hp up to an amazing 669 hp in the flagship T9.670.

As always, the new numbering system was gradually expanded across the whole tractor range and the T4000 tractors became the T4 line in 2011, with power ratings from 55 to 114 hp. Four-cylinder FPT engines provided the power and as usual several variants such as narrow and standard tread were offered.

Inevitably the T6000 line gave way to the T6 range, with seven models of four and six-cylinder tractors from 120 to 175 hp making up the Basildon-built range. As with the larger T7 line, a host of features and options was available to make these tractors as simple or as complex as required.

Last of the 'T' family to appear were the T5 tractors from Italy. Although not as sophisticated as the larger T6 tractors, these Italian-built machines still had a wide choice of options and features and included an all-new cab with a much larger area of glass.

With the whole range being upgraded, it seemed as though the New Holland tractor range would now settle down, but this was far from the case! New engine technology and the changing demands of the farming industry saw the whole range updated even further within just a few years.

The first T7 tractors arrived in 2010 and began a new numbering system for New Holland tractors with the range number first, followed by the maximum power output as a model number. Thus, the T7.200 was a T7 tractor with a rated power output of 155 hp and a maximum output of 203 hp.

The T7.210, such as this example shown preparing potato beds in Suffolk, was the most powerful of the standard build T7 models, the bigger models having a longer wheelbase.

Several T7 models could be specified with special bonnet decals including the British flag, denoting that they were British built. Even more of them carried the flag in between the two front lights on top of the cab.

A T7.220 of the longer wheelbase models waits for the snow to melt.

Biggest of the T7 range was the 269 hp T7.270 and a batch of these were painted in this special livery in 2014 to celebrate fifty years of the tractor factory in Basildon.

The special 'Basildon 50 Years' logo that was fitted to all the New Holland tractors built there in 2014.

The Blue Power versions of the T7 tractors proved very popular, with a full specification and metallic blue paint and chrome finish.

The T8 range also arrived in 2010 and these were radically different to the T8000 line they replaced, as shown by this flagship 389 hp T8.390.

By more closely following the lines of the Case IH Magnum it shares its platform with, the T8 could get more of its power to the ground when using multi-furrow ploughs, for example.

The sophisticated control centre in the cab of the T8 tractors.

With a maximum power rating of 502 hp, the T9.505 is no tiny tractor – but it was nowhere near as powerful as the flagship T9.670.

446 hp was available from the FPT six-cylinder engine fitted under the large bonnet of the T9.450.

A standard tread variant of the T4 range, the T4.95 was a 99 hp tractor.

The orchard version of the 107 hp tractor was the T4.105F. (Photograph: Sascha Jussen)

A T4.105F showing its ability to use implements on the front as well as rear linkage. (Photograph: Sascha Jussen)

All the New Holland range gradually included the new engine technology, including the T4.85.

The T5 line from Italy includes a new design of cab with rounded quarter sections. (Photograph: Kim Parks)

A T5.105 showing off the new cab.

The T6.175 was the largest of the new T6 range built at Basildon beside the T7 range.

CHAPTER 7

Latest Generation

All the current range of New Holland tractors meet the Tier 4B emissions regulations and, as each series of tractors was introduced, a new bonnet design arrived, allowing for extra cooling of the power units, which was an important consideration for the new emissions technology.

The T8 range, built at Racine in Wisconsin in the USA, now comprises five models from 320 to 435 hp, pushing the rigid tractor power envelope even further upwards. Metallic Blue Power versions of the T8 machines include the Auto Command constantly variable transmission but are limited to the T8.410 and T8.435 models only. A new development is the option of rubber track units instead of rear wheels on the T8 tractors. This option, known as SmartTrax, combines the manoeuvrability and flexibility of a conventional wheeled tractor with the larger footprint and less ground compaction of tracks.

The new T9 range consists of five models ranging from 469 hp up to a massive 692 hp flagship, making the T9.700 one of the most powerful production tractors in the world.

The new T7 tractors have changed a great deal with the introduction of Tier 4B compliant FPT engines and have been split into four standard wheelbase models from 175 to 225 hp, and four models in the long wheelbase range from 225 hp to 270 hp. FPT six-cylinder engines power all, including the two largest models in the T7 range, the Heavy Duty T7.290 and T7.315. With 313 hp available, the T7.315 is the largest tractor to ever be built at the Basildon factory and provides plenty of power in a fairly compact package which is considerably lighter than the T8 models. This sector of the market has been one that has only appeared fairly recently, with the idea of having much more power in what would have been a smaller mid-range size machine. That is not to say that the T7.290 and T7.315 are small; in fact, they are considerably larger than the smaller T7 models and have to be built on their own production line in the Basildon factory as they are too big to fit on the automated line that the T6 and the rest of the T7 range are assembled on.

New T5 and T6 tractors have also appeared, sharing the same new bonnet design of the T7 and T8 ranges with larger grilles. The 99 to 117 hp four-cylinder T5 models are built in Italy and the four- and six-cylinder T6 models from Basildon range from 125 to 180 hp with many transmission options, including the Auto Command, on the four largest models.

All these new models show that New Holland tractors are still up there as one of the world's leading tractor brands, and continue to sell to farmers and contractors around the globe.

New Tier 4B emissions compliant FPT engines gradually saw all the various New Holland tractor ranges updated over the last few years and also the inclusion of new styling with larger bonnet grilles. The T8 line started the trend with the new 435 hp T8.435 shown here.

A new option on the T8 tractors was the SmartTrax system, which saw the two rear wheels being replaced by a pair of rubber track units.

The SmartTrax option is only available on the three largest models and provides increased traction with a larger footprint area.

Inside the cab of the T8.435, the New Holland 'family look' of the control interfaces is obvious, with most functions being controlled from the armrest console.

The T9 range also received new FPT Cursor engines, taking the largest model up to 692 hp.

A range of Boomer compact tractors are built by New Holland in a purpose-built factory in the USA.

Blue Power versions of the latest T7 series tractors are still popular, as shown by this T7.225 of 225 hp, which is part of the long wheelbase T7 line-up.

Inside the cab of a T7.225 Blue Power, showing the Auto Comfort seat and padded passenger seat.

Once again all the main controls are positioned on the armrest in the T7.225 cab.

The T7.315 Heavy Duty tractor takes the power of the T7 range up to bigger tractor territory, with 313 hp available and a higher build than the smaller T7 tractors.

The T7.290 is the other Heavy Duty member of the T7 range with 290 hp available in the same chassis as the bigger 315 model.

A close-up of the bonnet of a standard wheelbase T7.245, showing the extra grilles.

The T7.315 looks stunning in the special Blue Power livery.

With its 313 hp rating, the T7.315 is the largest tractor built at the factory in Basildon and pushes into what had once been T8 territory.

The T6.180 is the only six-cylinder model offered in the T6 range.

Seen here in Blue Power livery, the T6.175 is the largest four-cylinder machine in the T6 range.

The T6.180 that, along with a T6.145, completed a tour around the coastline of the UK in 2017 for charity.

CHAPTER 8

The Future

Sophistication and ease of use are the two main watchwords for today's tractors, with computer and GPS technology making the driver's lot an increasingly stress-free one. New Holland tractors embrace all the latest satellite navigation and automation that is to be found on all modern tractors, but CNH has gone even further.

Back in 2009 the company showpieced a prototype hydrogen-powered T6000 series tractor that used a compressed hydrogen system to power an electric motor. Rated at over 100 hp, it was also equipped with a power take-off and an electrically driven constantly variable transmission.

More recently T6 tractors have appeared powered by methane gas, and during 2016 a Methane Power T6.180 tractor did a working tour of Europe as a proof of concept. With the increasing debate about harmful diesel fumes, could methane be the answer to future tractor power?

CNH has also been busy with autonomous tractor designs, using the Case IH Magnum and New Holland T8 platform to develop a driverless tractor. Significant trials in the USA appear to prove this technology as a viable form of farm power in the not too distant future and the Magnum version is not even fitted with a cab, although a small driving platform is fitted. Aspects of this driverless technology is now ready to be rolled out into the real world of agriculture, giving us an exciting glimpse into what will be the norm in years to come.

From the separate lineage of both Fiat and Ford tractors, the New Holland name has surfaced on the sides of tractors after being used on farm machinery for many decades previously. Three great names in agricultural mechanisation brought together under the umbrella of the Fiat organisation in 1991 have blossomed to provide a fully integrated and global presence in the sphere of tractors and farm machinery production, the likes of which would never have been imagined at the beginning of the 1990s. Whatever the future of farm mechanisation is, you can bet that the New Holland name will be prominently at the forefront of it.

Ever since 1895, the name of New Holland has been associated with farm machinery, and it looks like the future may well still be blue and yellow!

Based on a T6000 series tractor, New Holland built this hydrogen-powered tractor as a working concept.

With an output of 100 hp, this tractor was completely free of any polluting waste and was very silent in operation.

The CNH Industrial factory at Basildon in Essex, which is still home to the T6 and T7 range of tractors after fifty-four years of tractor assembly.

Freshly assembled New Holland tractors at the factory in Basildon awaiting collection for dispatch to all corners of the globe.

A contractor's New Holland TM and TSA tractors taking a break from carting silage in East Sussex.

New and old New Holland tractors cultivating at a working day in Wiltshire. The blue line is a long one and is still a favorite of many farmers and enthusiasts.

Two Blue Power New Holland tractors from the current T7 range at work at a demonstration day in Kent.

A Blue Power T7.315 powers away with a Kuhn plough on demonstration. Power has increased enormously over the years, as has computer technology. Not long ago a 313 hp tractor would have been the largest produced, but today there are two other more powerful ranges above this tractor in the New Holland product line.